Mathematical Based Stories and Observations by a Former Adjunct

Leonard Sperduto

CONTENTS

Preface

I am a former adjunct from Camden County College, Blackwood New Jersey. I taught in the basic mathematics department from 1992 to 1998.

This book is presented in two parts. The first part is a collection of mathematical based short stories. The second part are my observations on mathematical concepts from different areas of mathematics.

The Legend of Aunt Sally, The Land of Conic, Denom and Numer's Vegetable Garden, and *The Fibonacci Zoo* have been reprinted from *Short Stories and Essays.*

The Land of Conic's National Holiday and *Quadrilateral Tax* have been reprinted from *Trade with the Planets of the Tixe Games and Other Stories.*

Math Hell, A Triangular Love Story and Zero Thirteen are new stories.

The Mathematical Observations are my observations on how mathematics is represented in other areas as well as in the real world.

Future additions to this book will be published upon my writing more mathematical based short stories or finding more connections between math and the real world and other areas based upon my observations and self-study.

Leonard Sperduto

Mathematical Based Stories

THE LEGEND OF AUNT SALLY

Sally lived a very long time ago in a kingdom called Decagonia. Sally was the chief mathematician to the royal court. For many years, the people in the kingdom had no semblance of order when solving basic mathematical problems. A problem such as $11 + 13 \times 4 \div 10 - 12$ would have more than one answer. Sally decided to form a set a rules that would result in one specific answer for a specific problem.

Sally came up with the idea that multiplication and division should be computed before addition and subtraction. All multiplication and division should be done left to right before adding and subtracting. Addition and subtraction should be done at the end from left to right also. For the problem $11 + 13 \times 4 \div 10 - 12$, $13 \times 4 \div 10$ is calculated first which results in 5.2. The problem then becomes $11 + 5.2 - 12$, and the final result is 4.2. Sally presented this idea to the king, and he was very pleased.

The king made a declaration that all mathematical problems that have multiplication, division, addition, and subtraction in them should be done Sally's way. He named the method My Dear Aunt Sally. The king is the nephew of Sally. Sally was the youngest sister of the king's late father. As it stood now, the order of mathematical precedence is multiplication, division, addition and then subtraction. Sally added that multiplication and division should be done in the order that the two operations come in first, and then addition and subtraction should follow in the same manner.

The people of Decagonia were happy but still had some questions. The consensus was that this method if fine for addition, subtraction, multiplication, and division. The people then asked, "What happens when parentheses and powers appear in a problem?" The king presented the question to his aunt who then started work on the problem.

Sally decided to make powers a higher order than multiplication and division. She also claimed that all work should be done inside parentheses first following the order of precedence. She formally stated that all work should be done inside parentheses first with the inside of the inner most pair done first. Work must be done to remove all the parentheses from the inner most to the outer pair. Once all the parentheses are removed the operation is as follows: powers, followed by multiplication and division, and finally addition and subtraction. The precedence should take place inside every set of parentheses. To solve a problem such as $(11 + (11 \times (7 - 3) \div 2) - 4) \,^{\wedge}2$, $7 - 3$ is done first which results in 4 and the problem then becomes

$(11 + (11 \times 4 \div 2) - 4) \,^{\wedge}2$, followed by $11 \times 4 \div 2$ which results in 22 and the problem then becomes

$(11 + 22 - 4)^{\wedge}2$, followed by $11 + 22 - 4$ which results in 29 and the problem then becomes $29^{\wedge}2$ and the result is 841. Sally presented her work to the king, and he was pleased once again.

The king made a new declaration that all mathematical problems that have powers, parentheses, multiplication, division, addition, and subtraction in them should be done Sally's way. He named this method Pretty Please My Dear Aunt Sally. As it stood now, the order of mathematical precedence is work inside parentheses is to be done first and the precedence for the operations is powers, multiplication, division, addition and then subtraction. The king then had his aunt's work published and placed in the kingdom's library for all to use.

News of Sally's work spread to the nearby kingdoms. Kings and Queens sent their scholars to study Sally's method. Sally's method was gaining popularity. Scholars came from all over the world. There was an accepted order of mathematics. It looked like the entire world was happy with Sally's method.

Chaotica is located in one of the far corners of the globe. News of Sally's method took years to reach Chaotica. The dictator was not pleased. He rules his land with an iron fist. His subjects must obey him implicitly, or perish. He started formulating a plan to execute all who uses Sally's method. The dictator wanted everyone on the planet to solve problems his way, and no other.

Chaotica started invading the nearby lands. Kingdoms that did not abide by the dictator's rule were destroyed. This campaign lasted for years. Kingdoms were either enslaved or destroyed. The enslaved kingdoms reluctantly gave into the dictator's rule for fear of their lives. Chaotica finally marched into and destroyed Decagonia. Sally was burned to death along with her book. It looked like Chaos would rule the world

The dictator led his army back to Chaotica thinking he put an end to Sally's mathematical order. Someone should have checked to see if everyone was dead in Decagonia. One of the librarians was pinned under a bookcase. It looked like the bookcase had crushed the librarian to death. Somehow the librarian crawled out from under the bookcase. The librarian knew where to get help to restore mathematical order.

The librarian's ancestors were from the town of Relminica. Relminica was a small town that was surrounded by mountains. No one knew of its existence. Anyone who left kept the location a well-guarded secret. The town was created by a powerful wizard named Relmin. The town was named after him. One of the wizard's descendants still lived in the town. The librarian knew that the current wizard could help.

It was a long journey, but the librarian made it to Relminica. The librarian rested at her cousin's house before visiting the wizard's castle. The librarian made her way to the castle when she was strong enough to make the trip. The castle is on the side of one of the mountains.

The librarian made her plea to the wizard to restore mathematical order that Sally created. The wizard said, "I cannot restore mathematical order because I did not create it. Sally created mathematical order, and only she can restore it. I can resurrect Sally so that she may restore mathematical order." The wizard and librarian began making preparations for the trip to Decagonia.

The librarian and wizard made their way to Decagonia. The wizard was surprised to see all the destruction that the army of Chaotica made. The wizard and librarian rested at the librarian's house before performing the ritual that would bring Sally back to life. The distance from Relminica to Decagonia is quite long, and the wizard was too exhausted to perform the ritual.

The librarian and wizard went to the castle the next day to search for Sally's body. They found Sally's charred remains in her office. The wizard began to prepare for the resurrection ritual. He placed a blanket with ancient symbols over Sally. He lit scented candles. He sprayed the room with a strange scent. He sprinkled a deep black colored powder on top of the blanket. He kneeled at the covered body with his arms crossed against his chest. He chanted an incantation in an ancient language. The chanting continued for an hour. The wizard then rose and said, "It is done. We know must wait until the moon is full to see if the spell worked. Sally will come to me on the night of the full moon if the spell worked. We will wait at your house."

Three weeks later the moon was full. The wizard said, "Tonight is the night that Sally will rise from the dead if the spell worked." Four hours had passed and nothing happened. Suddenly the librarian's door quickly flung open and Sally was standing in the doorway. This was not the Sally of old. Sally had extensive burns from head to toe. Her tattered clothes barely covered her. The librarian helped Sally to a chair, and began to tell the story of the Siege of Chaotica.

Sally listened to the tale of how the people who followed her mathematical rules were either destroyed or enslaved by the people of Chaotica. Sally became angry. She wanted revenge. She pleaded with the wizard to help her destroy Chaotica and restore mathematical order. The wizard agreed. The wizard left the next day to gather everything he needed to resurrect an army for Sally. Sally rewrote her theory, and the librarian put the theory in book form in multiple copies.

Sally, the wizard, and the librarian started on their trek toward Chaotica. The trio stopped at each town between Decagonia and Chaotica. The librarian placed a copy of Sally's rules in each town's library. The wizard resurrected the dead from each town. All the resurrected citizens joined Sally on her quest. Mathematical order was being restored. The final battle had yet to be fought.

Sally and her army arrived at the gates of Chaotica. Chaotica's dictator was surprised. The dictator stated, "I destroyed you once before and I will do so again." The final battle for mathematical supremacy had begun. Chaotica's army was slowly being depleted. Sally's army went untouched. How could the army of Chaotica destroy what was once dead?

As the battle raged on, Sally, the wizard, and the librarian made their way to the dictator's strong hold. The trio confronted the dictator. The dictator was defiant. Sally asked for the dictator to surrender. "I shall never surrender, "said the dictator. Sally lunged at the dictator. Her hands tightly griped the dictator's throat. The dictator tried to yell for help, but the wizard put a protection spell around the door so that no one may enter. With his last breathe the dictator gasped, "You won Sally. I now know the error of my ways." The dictator died still in Sally's grasp.

Sally released the dictator. The librarian placed the final copy of Sally's rules on the remains of the dictator. It was a symbolic gesture to claim that mathematical order was restored. Suddenly the battle stopped. The citizens of Chaotica were awakening from an evil spell. It turned out the dictator was an evil wizard who put a spell on the population of Chaotica. The spell was broken with the dictator's death.

Sally was pleased that mathematical order was restored. A light appeared out of nowhere and bathed her. She felt at peace. Her body began to heal. After a few moments, the light and Sally were gone. The librarian asked the wizard, "What happened to Sally's body?" The wizard replied, "Sally is at peace. She has moved on to a higher plane of existence. She will be watching to see her rules followed for all eternity now. It is time for all of us to go home now."

THE LAND OF CONIC

Hello. Welcome to the land of Conic. My name is Sir Cul. I am a knight for King Xaxis, and I will be your tour guide through the land of Conic.

Our land is divided by the Yaxis River. The land on both sides of the river forms a perfect hyperbola. The west side of the river is our residential section, and the east side is our industrial section. Our citizens must cross the Transverse Bridge located at the center of the land.

The center of the Transverse Bridge is exactly over the center of the river. Each end of the bridge is of equidistant length from the center. We will go over the bridge later in our tour.

Let us begin our tour. You are at the west gate to the land of Conic. You would have met the king's daughter, Princess Directrix, if you entered our land at the east gate. She is the other tour guide. This is our residential section where most of our citizens work in factories on the other side of the Yaxis River. There is a small population that does own shops on this side of the river. The shops are of the retail variety, where the people by their household goods.

You will notice that all our homes are of a standard format found in every part of the world. Our TV antennas are of a parabolic shape that makes them more powerful than standard roof top antennas. Our citizens enjoy programs not only from our station, WPVF, but from stations from the surrounding lands.

Please enter the tour bus. We will travel down Axis Avenue which is the route to the Transverse Bridge. On our way back, we will stop and visit the shops along Focus Street. We may get to see the Princess' tour bus if she is conducting a tour. Her route is the exact opposite of mine.

Here we are at the Transverse Bridge. It is a simple drawbridge. Carl and his daughter Nancy collect the tolls for the bridge. Carl also raises the bridge to let the ships pass safely by. They both work from six in the morning to six in the evening. The bridge is closed during the night since our factories only operate during the day. The bridge is left in the raised position at night.

"Good afternoon Sir Cul. Do you have your bridge pass?"

"Yes I do Carl."

"Fine. There won't be any ships coming through today. You will be able to keep your schedule."

"That's good. This group won't be complaining about hunger when we get back to Focus Street. Enjoy the rest of your day."

"Thank you. Enjoy the rest of your day Sir Cul."

Here we are in the industrial section. The first few buildings are our food processing plants. We will stop at the juice plant where everyone will receive a free bottle of juice and ice cooler on the way back.

This next section is where our small appliances, clothing and sporting goods are manufactured. These factories ship their goods to the shops along Focus Street. They also export their goods to the surrounding kingdoms.

The final section houses the factories that make our furniture, larger appliances, and fixtures for our bathrooms and kitchens. These factories have showrooms where our citizens decide on what to buy. The factories ship and install the purchases within two days to the purchaser.

This ends our industrial part of the tour. We will proceed to the east gate and then turn around. We may get to meet the princess at the gate.

Here we are at the east gate. The princess is not here, but Sir Cumference is. Sir Cumference is my brother knight.

"Hello, fellow knight. Why are you here instead of the princess?"

"Hello, Sir Cul. I am filling in for the princess for the next few days. She is preparing for her trip to Decagonia to attend the feast of Aunt Sally."

"I had forgotten that the feast was only a few days away. The princess attends every year to honor the woman that defeated chaos and restored mathematical order to the world."

"Do you have a tour today Sir Cumference?"

"Nay, Sir Cul. I'm just guarding the gate and checking the trucks that are arriving and departing today. It is just a routine day."

"How goes your tour?"

"Fine, we are on schedule. We are heading back to the juice factory."

"Then let me keep you no longer my brother. May the remainder of your tour continue to be pleasant."

"Thank you, brother knight. Enjoy the rest of your day."

"Thank you I will."

We have arrived at the juice factory where you will choose one bottle of any flavor that you desire along with an ice cooler. The ice coolers have one of our kingdom's symbols on them. You can choose a cooler with a circle, ellipse, hyperbola or parabola on it. We knights are partial to the one with the circle on it. The stop at the juice factory will be forty-five minutes.

I hope all of you will enjoy your juice and coolers. We will return over the Transverse Bridge and head to Focus Street where everyone will have a two hour shopping spree.

Here we are at the Transverse Bridge again.

"Good afternoon Sir Cul. Do you have your bridge pass?"

"Yes I do Nancy."

"Fine. How is my father's mood today?"

"He was pleasant. Why do you ask?"

"Yesterday he was complaining about some teenage picnickers on the river bank near his station. They were playing their music too loud. He complained about the boom-boom music all the way home."

"That's your father. Any music he doesn't like he calls boom-boom music."

"I know. Hopefully, there won't be any picnickers today playing anything that he doesn't like. I just want a quiet trip on the way home tonight. I love my father dearly, but he can be a bit annoying when he complains about unimportant matters."

"Enjoy the rest of your tour Sir Cul, and have a peaceful evening."

"Thank you Nancy and you have a peaceful evening also."

We have arrived at Focus Street. You have two hours to shop and eat at any of the restaurants. The gift shop has ashtrays shaped like ellipses, and posters of our kingdoms' symbols. The book store has books on the origins of our symbols. Any of the restaurants are good, and have traditional conic cuisine.

My personal favorite place is Amy's Bakery. She uses the same fruit as the juice factory to fill her pies. Her cookies are perfect circles and ellipses. Her best dessert is a cone made from yellow cake mix covered with her special chocolate sauce and topped with either marshmallows or sprinkles.

Please enjoy your time on Focus Street. I will see everyone in two hours.

Welcome back to the tour bus. I hope everyone enjoyed Focus Street. Let me take a count to make sure everyone is here. I remember one time when a passenger did not want to leave Amy's bakery. He was addicted to the smells, and did not want to leave. He is now a resident, and spends almost every day at the bakery.

Everyone is here and now let's return to the west gate.

We are back at the west gate. Thank you for taking the tour. I hope you enjoyed your time here in the Land of Conic. Please visit us again, and have a safe and pleasant journey back home.

DENOM AND NUMER'S VEGETABLE GARDEN

A farmer decided to have his two children grow their own vegetable garden so that they can learn to be farmers. He called them and said, "It is time for you two to start growing your own vegetables. I will give you an area of land which will be a multiple of the number of types of vegetables that you grow. Let's say that you decide to grow three different types of vegetables. I will give you an area that is a multiple of three. Now go and decide on what types of vegetables that you will grow."

Denom and Numer went into their room and planned out their garden. They decided to plant vegetables that grow in the ground since they both like playing with dirt. They decided to grow five types of vegetables. They went back to their father and told him their decision.

The farmer said, "I will give you an area that is a multiple of five. It us up to both of you to figure out how much land that you need."

Denom and Numer returned to their room to plan their garden.

Denom said, "We could make it a square garden which means that we will need twenty-five square feet of land."

"Yes, but we should figure out the size that we will need for each vegetable. We may not want the same size area for each one."

Denom agreed and they looked at their list of vegetables. Carrots, horseradish, onions, potatoes, and radishes were the vegetables that the two children wanted to grow.

"We all like onions and radishes in our salads," said Numer. "We should give them equal space."

Denom agreed. "Papa loves carrots. Carrots should have the largest space."

Numer agreed. They continued to look at their list. They both realized that their grandfather liked horseradish, but they do not see him that often. So they decided to make the area for horseradish the smallest. They will have a nice amount to bring to him the next time they visit.

They also decided to give potatoes a large area since their mother can make different dishes with potatoes. Both children had visions of French fries, scalloped potatoes, baked potatoes, potatoes and eggs, and potato pancakes for future meals.

"Now let's see what we have," said Denom. "We are giving carrots and potatoes the largest areas, horseradish the smallest area, and onions and radishes will be the middle area."

"Let's give potatoes and carrots two-fifths each. We'll give radishes and onions half of the area of potatoes and carrots. Horseradish will be half the size of radishes and onions."

"That will make potatoes and carrots two-fifths each, radishes and onions one-fifth each and horseradish one-tenth."

"Wait a minute. Your figures add up to thirteen-tenths. Two-fifths plus two-fifths plus one-fifth plus one-fifth plus one-tenth equals thirteen-tenths. I see your mistake. You divided the two-fifths by two to get one-fifth, and then divided the one-fifth by two to get one-tenth. You forgot that radishes, onions, and horseradish must equal one-fifth since potatoes and carrots make up four-fifths of the garden."

"Let's ask papa for help. He used to teach math. Neither one of us can figure out how to make three types of vegetables fit into one-fifth if the areas for the vegetables are not all equal in size."

They presented their father with their figures. The farmer knew immediately what the children did wrong. He explained that algebra was needed to solve the problem. He told both of them to go and play until he finishes the calculations.

The farmer started working out the problem. Radishes and onions are equal size, but horseradish was half the size of radishes or onions. The area for the three types of vegetables must equal one-fifth. Let x equal the size for the radishes and onions and one-half x equal the size for the horseradish then $x + x + \frac{1}{2}x = \frac{1}{5}$. Simplifying the equation means that $\frac{5}{2}x = \frac{1}{5}$. The solution to the equation is two twenty-fifths which means that onions and radishes get two twenty-fifths each of the one-fifth, and horseradish gets a twenty-fifth of the one-fifth.

Numer calculated the figures and found out that everything now adds up. He explained to his brother that the garden needs to be separated into twenty-fifths. Ten twenty-fifths of the garden will belong to carrots and another ten twenty-fifths will belong to potatoes. Two twenty-fifths of the garden will belong to onions and another two twenty-fifths will belong to radishes. Horseradish will belong to the remaining twenty-fifth.

Numer said, "Let's look at our original area. We said that we will have a garden of twenty-five square feet. Ten twenty-fifths of twenty-five square feet is ten square feet. Two twenty-fifths of twenty-five square feet will be two square feet. One twenty-fifth of twenty-five square feet will be one square foot. Each section will be kind of small."

"Let's increase the area to one-hundred square feet and see how each section grows."

Numer recalculated the smaller areas and came up with ten twenty-fifths of one-hundred square feet is forty square feet. Two twenty-fifths of one-hundred square feet will be eight square feet. One twenty-fifth of one-hundred square feet will be four square feet.

"This is fine. Let's tell papa that we need one-hundred square feet of land."

The farmer was pleased with his sons' plan. "You will have one-hundred square feet of land. You may start working on your garden tomorrow."

Numer and Denom were happy. For the rest of the day, they put together the tools that will be needed for the morning's labor. They went to sleep with smiles on their faces. They both looked forward to starting their garden.

The farmer called his children and explained that they needed to give onions and radishes two twenty-fifths, and horseradish one twenty-fifth of the garden.

The next day, the brothers woke with excitement. They rushed their breakfast and immediately got their tools and went to the area in which their father laid out for them.

Numer said, "Before we start digging, let's separate the garden into the areas that we specified."

The brothers marked off two areas of forty square feet, two areas of eight square feet and one area of four square feet. They were ready to start digging and planting.

"Let's start with horseradish first since it's the smallest area," said Denom.

Numer agreed. "Then let's do the radishes and onions next. We'll save the potatoes and carrots for last."

Denom agreed and they started digging and planting their garden.

The two young farmers spent all day in the garden. They did take a break for lunch. They finished their work by suppertime. They were exhausted. Normally they would watch their favorite television shows before going to bed. Tonight, they were only able to watch one show.

Every day, Denom and Numer tended their garden. They watered it when there was no rain. They pulled out the weeds so the weeds could not strangle their plants. Eventually the two young farmers were able to start picking vegetables. Some carrots, onions and potatoes were part of their first harvest. The radishes and horseradish were not ready to be picked. The brothers did not mind. They knew it would only be a matter of time when they could pick every vegetable.

They presented their first harvest to their mother. "What a nice crop you have there. I will use the onions for tonight's salad and I will make scalloped potatoes for you tomorrow. I will also make you a carrot cake tomorrow to reward your hard work".

"But mama," the brothers said in unison. "The carrots are for papa."

"Don't worry about papa. He won't mind if the first batch of carrots goes into a cake to reward you two. Now go and wash up. Supper is almost ready."

During supper, the farmer told his sons how proud he was of them. He also told them that he doesn't mind waiting for the next batch of carrots. He can wait for his carrots for snacks.

Denom and Numer were happy. Both their parents were proud of them. They went to bed with big smiles on their faces.

The next day, the two young farmers went back to their routine of watering and weeding their garden. They enjoyed their scalloped potatoes and cake at supper.

As time went on, more vegetables were picked. Salads were always full of fresh onions and radishes. Mama always changed the way that she made potatoes. One day there would be mashed potatoes, another would be baked, and on Fridays she would make French Fries. She also made potato pancakes every now and then.

Papa enjoyed snacking on the carrots. Sometimes the carrots were added into the salad. Every now and then, boiled carrots were served with supper.

At the end of the season, Denom and Numer gathered up their remaining harvest. For some reason, the horseradish took the longest to grow but they did not mind since they were able to give their grandfather the entire crop of horseradish.

The two young farmers were very happy with their first garden. Both their parents and their grandfather were proud of them. They hoped that next year will be as successful as this year.

THE FIBONACCI ZOO

Welcome to the Fibonacci Zoo on the planet Cusaba. I hope your experience through the portal was a pleasant one. I am robot F55, and I will be your guide through the zoo. I am programmed to speak in all languages in the known universe, and that is why you are able to understand me.

Let me tell you a little about the zoo before our tour. We have a collection of creatures from different times, planets and dimensions. Each species is encased in a dome that is equivalent to their original environment. There are three domes in each area that we call a tripod that we will visit. The goal of the zookeeper is to have the zoo cover every inch of the planet. But this goal is still a long way to completion. We only recently started to conduct tours. The zookeeper wanted to be sure that the portals and space ports were safe for our visitors before beginning operations.

You may now enter the hover bus. The atmosphere inside the bus is now equivalent to the one from your planet. It takes a while for each bus to become atmospherically correct before beginning our tour. This is why we give our little tale about the zoo while our visitors wait in the holding lounge.

Our first stop will be the rare animal tripod. You will see creatures that will not be found in any other zoo. Only the zookeeper has the technology to bring these creatures here.

A centaur resides in the first dome. He comes from one the mythological dimensions of the planet Earth. From the waist up, he has the form of a human male. From the waist down, he has the body of a horse. The centaur is a warrior at heart and he constantly tries to break the dome to attack either the dragon or the unicorns. The domes are impenetrable, but the centaur keeps trying to break through. He must have recently attacked the dome. As you can see, he is resting now. He always rest between attacks.

A dragon lives in the second dome in this area. It is also from one of Earth's mythological dimensions. Dragon's would terrorize the country side by breathing fire on anything that lied in their path. Many dragons were slain by knights. Our dragon is safe from harm, but it tries to escape by breathing fire on the dome. As you can see, it is happily flying around now.

Two unicorns from the planet Gonaldia reside in the third dome. These unicorns are almost similar to the ones found in one of Earth's mythological dimensions. They look like a horse with a horn growing out of their forehead, but the main difference is in the horns. When two Gonaldia unicorns touch their horns, a spectacular fireworks display occurs. The display is different every time. Let me radio Jim so that he can darken the unicorn's dome so that we might be able to see the fireworks. Jim is the head of environmental controls. He makes sure that each dome is working properly.

The dome is starting to darken. It will only be a matter of moments before we see the fireworks. The display has started. I haven't seen that shade of green in quite a while.

We have replicas of the horns in our gift shop. A group of earthlings from The Land of Conic were fascinated by the conic shape of the horns. They wanted to buy a replica, but at the time we did not have any. The zookeeper designed a replica and now when the people from the Land of Conic visit, they buy the horns.

Our next stop is the waterfowl tripod. Two of the three domes house water fowls from the planet Earth. The first one is home to a family of ducks. As you can see there is a family of three swimming in their pond. Earthlings have a tendency to hunt ducks for sport and food. These three are safe from harm.

The next dome houses a family of five swans. Swans are beautiful white birds with long necks. The swan has been portrayed in ballet, and swan replicas made out of tin foil have been used as centerpieces for Earth celebrations. Swans are one of the more revered water fowls on the planet Earth.

The last dome in this section houses eight storks from the planet Alequ. These storks are exactly like the ones found on the planet Earth, but these storks do deliver babies. Babies on Alequ are chemically engineered based on their parents' request. A stork delivers a baby to the parents' home once the baby has emerged from a chemical pool. The eight storks that we have here are retired. These storks are too old and weak to deliver babies.

Our next stop is the aviary tripod. Thirteen mockingbirds from the planet Nichero reside in the first dome. These black birds look exactly like crows from the planet Earth. The Nicheroian mockingbirds imitate any sound in unison. We are fortunate to arrive here at this time because in one minute a random sound will engulf the dome and the mockingbirds will imitate the sound. Every fifteen minutes a random sound engulfs the dome. I wonder what sound the birds will imitate. The birds are imitating a fog horn from the planet Earth. Not one of the more pleasant sounds.

Thirty-four doves from the planet Earth reside in the second dome. A dove is a small white bird. They have been known to be used in Earth ceremonies such as weddings and opening days at special events. The dove is also a symbol of peace on the planet Earth.

The last dome houses fifty-five owls from the planet Earth. The owls mostly sleep in the day and fly at night. The owl is supposed to be the wisest of the birds on the planet Earth. Owls have been depicted as academic scholars in many forms of Earth media. Judge for yourself as to whether the owl has any intelligence at all. Many of our visitors have debates about the owl's intelligence.

Insects live in the next tripod. The first dome houses eighty-nine beetles from the planet Danjor. These beetles look exactly like Earth beetles with one major exception. The Danjoran beetles are one thousand times the size of their Earth counterparts. The Danjoran beetles are quite tame on their own planet, but they cause havoc on any other planet. They tend to migrate to other worlds in the Danjoran solar system every eleven Danjoran years and cause devastation. No one fully understands the behavior of the Dajoran beetle. There is only insect life with insect intelligence on Danjor. The migration was observed from scientists from a neighboring solar system using a powerful telescope. These scientists are still working on a behavioral theory on the Danjoran beetle.

One hundred forty-four spiders from the planet Lejou live in the second dome. These spiders are fifty times the size of their Earth counterparts. The webs from the Lejouan spider are used for making winter clothes. The winters on the planet Lejou become extremely cold with temperatures far below the planet's freezing point. The web is an unusual thermal material that keeps the bitter cold temperatures away from the wearer's body. Many cold planets have been known to trade with Lejou for the web material.

The final dome houses two hundred thirty-three butterflies from the planet Earth. Butterflies come in different varieties. The main distinction is the different colors on their wings. Humans have been known to collect butterflies. Butterflies are usually mounted in a frame and displayed on a wall. The butterflies in the dome are free from harm, and live in the manner in which they are accustomed. On the weekends, an Earth woman known as Miss Butterfly gives lectures on the butterfly. The same technology that allows me to communicate with every known species allows our visitors to listen to her lectures.

The last tripod houses cute little creatures. Three hundred seventy-seven rabbits from the planet Earth live in the first dome. Rabbits have been known to live among humans as pets. Rabbits have also been used by magicians in magic acts. Rabbits are also a food source for humans. There is a human woman named Doctor Thump who gives lectures on the Earth rabbits on the days that she in not treating all our creatures. Doctor Thump is our veterinarian who takes care of every creature with great care and love.

Here is an interesting fact about rabbits. Rabbits tend to reproduce at rapid rates. The zookeeper sends the excess rabbits back to Earth to keep the population at three hundred seventy-seven. Rabbits may be moved to later tripods when more of the zoo is completed. As you may have notice, populations have increased after the second dome in the first tripod.

The second dome in this tripod houses six-hundred ten koala bears from the planet Earth. The koala bear hails from a country known as Australia. The bears live among and eat off of the eucalyptus trees as you can see. There is some speculation about whether or not the koalas can speak. We have video files that show a koala speaking about something called Qantas, but our koala bears do not speak. It is not known at this time what Qantas is since our data banks are incomplete. Our data banks are continuously being updated but for some unknown reason the data from the planet Earth is slow in reaching us. It may be because their technology is not that compatible with ours. The zookeeper is trying to rectify this problem.

Nine hundred eighty-seven bellooms from the planet Flora live in the last dome. Bellooms are flower like creatures that live in a forest. The trees that make up the forest are not typical looking trees. The trees do not have branches, but have long needles. The trees resemble cones. As you can see, the bellooms dance around in trios spreading their petals along the ground.

Unfortunately the cute creatures' tripod is the last one to visit. The remainder of the zoo is still under construction. The next tripod will hold one thousand five hundred ninety-seven, two thousand five hundred eighty-four, and four thousand one hundred eighty-one creatures in each dome. The types of creatures are not known at this time. We will return to your space portal after a brief stop.

As you can see that our hospital is right next to our gift shop. Dr. Thump is the head of the hospital. She must be busy preparing for her rounds since we did not see her in any of the domes.

The atmosphere inside the gift shop is now suitable to your needs. You have your equivalent of thirty earth minutes to shop.

Hopefully you enjoyed your visit to our zoo. You may see different creatures on your next visit and at least one more tripod will have been completed. The zookeeper does change some of the domes every so often. Nine of our lunar cycles ago, a Cyclops was housed where the centaur now lives. You may see a different fireworks display if the unicorns are still here.

It has been a pleasure to be your tour guide. Please come back soon and bring your friends and relatives. This is robot F55 signing off.

THE LAND OF CONIC'S NATIONAL HOLIDAY

Hello. Welcome to the land of Conic. My name is Sir Cul. I am a knight for King Xaxis, and I am your tour guide through the land of Conic.

Today is our national holiday. In honor of Pi Day, the price for the tour today is three conical dollars and fourteen conical cents.

The Yaxis River divides our land. The land on both sides of the river forms a perfect hyperbola. The west side of the river is our residential section, and the east side is our industrial section. Our citizens must cross the Transverse Bridge located at the center of the land.

The center of the Transverse Bridge is exactly over the center of the river. Each end of the bridge is of equidistant length from the center. The first half of the tour ends at the bridge since the industrial section is closed for the holiday.

Let us begin our tour. You are at the west gate to the land of Conic. A fellow knight leads tours from the east gate today. Normally the king's daughter, Princess Directrix, conducts tours from the east gate. The princess attends festivities all day today. This is our residential section where most of our citizens work in factories on the other side of the Yaxis River. There is a small population that does own shops on this side of the river. The shops are of the retail variety, where the people by their household goods.

You will notice that all our homes are of a standard format found in every part of the world. Each home has decorative circles in the windows today. The circumference and area of each circle is noted inside each circle with the measurement in terms of pi. Our TV antennas are of a parabolic shape that makes them more powerful than standard roof top antennas. Our citizens enjoy programs not only from our station, WPVF, but from stations from the surrounding lands.

Please enter the tour bus. We will travel down Axis Avenue which is the route to the Transverse Bridge. On our way back, we will stop and visit the shops along Focus Street. We may get to see the other tour bus. The other tour bus starts at the east gate and the route is the exact opposite of mine.

Here we are at the Transverse Bridge. It is a simple drawbridge. Carl and his daughter Nancy collect the tolls for the bridge. Carl also raises the bridge to let the ships pass safely by. They both work from six in the morning to six in the evening. The bridge is closed during the night since our factories only operate during the day. The bridge is left in the raised position at night.

We won't cross the bridge but will follow the path that parallels the bridge to observe how some of our citizens celebrate the holiday.

Notice that some of our citizens have picnics along the river bank. The traditional picnic consists of circular sandwiches, pies and cakes. Each picnicker measures the circumference and area of each sandwich, pie and cake before eating. Some of the school children bring their math books to the picnic to study circles.

We will now travel to Focus Street. You will have two hours to shop and eat at any of the restaurants. The gift shop has ashtrays shaped like ellipses, and posters of our kingdoms' symbols. The book store has books on the origins of our symbols. We have excellent restaurants that serve traditional conic cuisine. Every store and restaurant discounted the circular items for sale.

My personal favorite place is Amy's Bakery. Her cookies are perfect circles and ellipses. Her best dessert is a cone made from yellow cake mix covered with her special chocolate sauce and topped with either marshmallows or sprinkles.

We have arrived on Focus Street. Please enjoy your time here. Return to the bus in two hours and then we will go to the town circle to enjoy the afternoon concert.

Welcome back to the tour bus. I hope everyone enjoyed Focus Street. Let me take a count to make sure everyone is here. I remember one time when a passenger stayed in Amy's bakery. He loved the smells, and wanted to stay in the shop. He is now a resident, and spends almost every day at the bakery.

Everyone is here and now let's proceed to the town circle to enjoy the concert.

Here we are at the town circle. My brother knight welcomes Sir Cumference welcomes us.

"Hello brother knight. "You are not taking the princess' place on the east gate tour bus today?"

"Nay brother. The princess cancelled the east gate tour bus today. You have the only tour today. I am here to hand out the concert program to your passengers."

"Will the princess be attending the concert?"

"She will attend this evening's concert. She visits the hospital to celebrate the holiday with the sick."

"Tis a shame that my passengers will not meet our beloved princess. Please come on board and give my passengers their programs."

"Thank you and here is one for you my brother."

"The concert starts in thirty minutes," Sir Cumference said while exiting the bus.

"Let's see what program our orchestra planned for today. 'Circle Man,' 'What's my Area,' 'Pi or 3.14,' 'Circle This,' 'If You Want Pi (You've Got It),' 'Two Circles Have I,' and 'Tangent Time,'" Sir Cul said to his passengers.

Please find a seat in the concert area. We will return to the west gate after the concert. Please enjoy the performance.

I hope you enjoyed the concert. We will now return to the west gate. You will receive a commemorative pin in recognition of Pi Day upon disembarking the bus.

We have returned to the west gate. Thank you for taking the tour. I hope you enjoyed your time here in the Land of Conic. Please visit us again, and have a safe and pleasant journey back home.

QUADRILATERAL TAX

I decided to return to the Land of Conic. I enjoyed my previous trip. A whole land based on the conic sections amazed me. I decided to visit the town library instead of taking the tour.

I browsed through books about the history of the land. I learned that an ancestor of King Xaxis, who had a love of the conic sections, founded the land. Rivers and prominent streets were named after conic terminology.

I decided to browse through the tax books next. I found the usual taxes. Property, sales, and wage taxes made up the majority of the taxes. I found one tax most unusual. The Land of Conic had a tax called the Quadrilateral Tax.

One section of the code for the Quadrilateral Tax stated that residents who have any form of a quadrilateral on their property must pay the tax. Another section stated that any manufacturer who places any type of quadrilateral on any product must pay the tax.

I wondered if the Quadrilateral Tax generated any income for the Land of Conic. I never saw any windows in any other shape other than circles or semicircles. Residents who owned televisions would pay the tax if the televisions are cubic in nature or a rectangular flat screen. Manufacturers of the televisions would pay the tax as well. Manufacturers might make conical televisions instead of cubical or rectangular ones. I do not know since I never visited a residence nor a television manufacturer.

I wondered if the creators of the tax wanted to generate income or deter any form other than conical forms on the land.

MATH HELL

The Devil created a new demon. The Devil ordered the demon to create new tortures.

"All sinners who hate math will spend eternity in a mathematical concept, my lord," said the demon.

"Go and wait the arrival of these sinners," replied the Devil.

The demon bowed and left.

The demon arrived at the reception area and introduced himself to the demon for new arrivals.

"His lordship has become bored with the tortures. Every millennia he creates a new demon for new tortures. I'll will summon you if one of the arrivals meets your criteria."

The new demon bowed and left.

A new arrival met the new demon's criteria. The new demon came when summoned and placed a finger on the sinners head. The demon learned that the sinner hated trigonometry and had a fear of heights.

The demon and sinner vanished and reappeared at a roller coaster. "You will ride this roller coaster for all eternity," the demon told the sinner.

The sinner entered the car. The car rose a small distance then dropped some distance. The car then rose a larger distance and then dropped a larger distance than the previous distance. The progression continued. The distances became larger for every incline and decline. The roller coaster was based on the formula $f(x) = x \cos x$ where $x \geq 0$.

Another sinner arrived that met the demon's criteria. This sinner hated geometry and was claustrophobic.

The demon placed the sinner in a triangular room. The walls formed an equilateral triangle. "You will spend all eternity in this room."

The sinner thought nothing of the punishment and the claustrophobia could not occur. The walls shrank equally causing the interior to become smaller. The walls were about to surround the sinner when the claustrophobia occurred. The walls returned to their original position and the process repeated.

A third sinner arrived who hated geometry and had a fear of water.

The demon placed the sinner in a conical shaped tank with the point on the bottom and a circular lid on top. Water filled the tank. The sinner struggled to stay afloat. The sinner rose with the water. The sinner's head pressed against the lid. The water receded. The process repeated when the tank was completely empty.

The Devil summoned the demon. "I'm pleased with your work. Continue to use fear with math. I enjoy watching the fear on the faces."

The demon bowed, departed and waited for the next sinner.

A TRIANGULAR LOVE STORY

Dennis' mother gave birth to a son in a cab from the Triangle Cab Company. She gave birth a few blocks from the hospital.

Dennis was an average student through school except for certain concepts of mathematics. He excelled at triangles during mathematics. Dennis even took up the triangle for the stage band in high school.

Dennis attended college after high school. He studied architecture since he did well in Geometry and Trigonometry. He found employment after graduating from college.

Dennis' job bored him. The company always rejected his triangular designs. He designed standard buildings but wasn't pleased with his work. In his spare time, he designed a triangular home to live in since living in a rectangular apartment bored him.

Dennis hired a contractor once he had enough money saved. The contactor thought the house was unusual since it was a one floor building with a basement but every room was triangular within a triangular frame. Even the windows and doors were triangular.

The plumber thought that Dennis was somewhat of an eccentric since a triangular bathtub, and sink was placed inside a triangular bathroom. Only the toilet was standard.

The house was built on a triangular lot surrounded by three streets. Dennis found the lot while driving around town on vacation looking for triangular land.

Dennis moved in once construction was completed. He was happy at home but still bored at work. He knew that no one would want his triangular designs so he couldn't open up his own firm. Dennis decided to suffer in silence at work.

Dennis met a woman at a restaurant five years after the house was built. Dennis was leaving the restaurant when he spotted a woman making triangles with her napkin. He introduced himself and she allowed him to sit with her. They talked for a while and found that they both had similar obsessions with triangles. They made a date for the following week.

Dennis and Elizabeth went on a simple walk through a local park. They decided to make a triangular path. They talked about their love of triangles. Dennis told her about his house. Elizabeth said she would like to see it if the two of them became serious. Dennis agreed to the terms and they decided to have a picnic the next week.

The picnic was at the same park. Elizabeth rolled out a triangular blanket. The sandwiches were cut into triangles. The picnic basket was also triangular. The only items that weren't triangular were the ice tea bottles.

Dennis asked Elizabeth about the picnic basket. She said that she made it herself since she could not find one. The conversation turned to clothes. Both agreed that clothes were boring unless printed with triangles. Both wore ordinary looking clothes for both dates, but decided to wear triangle print clothes whenever they were together.

The picnic became a weekly event. Both wore clothes with triangle prints. Both started having strong feelings for each other.

"I'm starting to have strong feelings towards you and I would like to see your house," Elizabeth said after their recent picnic.

"I feel the same way. How about instead of a picnic next week, I make dinner for us at my house."

"I would love that."

"Do you want me to pick you up?"

"No, I'll drive. I don't want to interrupt your cooking."

Dennis gave his address to Elizabeth and made the date for 6'oclock.

Elizabeth arrived promptly at six.

"I like the outside of the house. How did you find the lot to build the house?"

"I was riding around town looking for a lot like this when I was on vacation. This seemed to be the only lot like it in town. Come in and I'll show you the inside."

The house impressed Elizabeth.

"I'm impressed. It's a shame that no one will buy your triangular designs. You have an amazing talent for triangles."

"Thank you. I wanted a place to live that suited me."

"I know what you mean. I'm tired of living in my rectangular apartment."

Thoughts of marriage started to form in Dennis' mind.

Dinner was a simple affair. Dennis made hamburgers in the shape of triangles. The rolls were cut into triangles. Dennis placed three French fries into a right triangle on Elizabeth's plate.

"You are really into triangles. I think I love you," said Elizabeth.

"I think I love you," Dennis replied.

"Let's just enjoy dinner and see what the future holds for us," said Elizabeth.

They finished dinner and Elizabeth kissed Dennis goodbye. Next week they would return to their picnic.

Elizabeth and Dennis talked about how much they enjoyed each other's company at his house at the next picnic. They agreed to have dinner at Dennis' house once a month.

Elizabeth and Dennis continued to date for another year with picnics three times a month and the monthly dinner at his house. Dennis decided to propose to Elizabeth. He decided to visit a jeweler to see if he could buy a diamond ring with the diamond inside a triangular setting.

The jeweler said that a triangular setting must be custom made.

Dennis ordered the ring.

The jeweler contacted Dennis a month later. Dennis picked up the ring and was pleased, and would propose at the next picnic.

Elizabeth accepted the proposal. They decided to have the wedding in the park next year. They decided on June 10. Elizabeth always dreamt of a June wedding. Six, the number for the month of June, and ten are triangular numbers which made the perfect date.

Elizabeth and Dennis made plans for a triangular wedding during their engagement. The chairs for the bride's and groom's sections would be aligned in triangles. Instead of a traditional tier weeding cake, the couple decided on a sheet cake with triangular slices cut. All guest will sit on triangular blankets during the reception.

Elizabeth's and Dennis' love grew. Both of them could not wait until the wedding day.

The wedding came. It was a perfect day. Most of the guests thought that it was odd that the wedding had a triangular theme. Everyone was happy that Elizabeth and Dennis found each other and were in love.

Dennis carried Elizabeth over the threshold at their, formally Dennis', house. They decided not to go on a honeymoon since they could not find one that suited their triangular interest. They decided that they will continue to have a picnic every week in the park.

Elizabeth became pregnant a year later. They decided not to have any more children. Three was a nice triangular number. The next triangular number is six, but neither wanted four children.

Elizabeth gave birth to a son. Elizabeth and Dennis named him Pythagoras after the mathematician who developed $a^2 + b^2 = c^2$ which is the formula for the squares of two sides of a right triangle added together equals the hypotenuse squared.

ZERO THIRTEEN

To a baseball player, 0-13 means that he was hitless for thirteen at bats.

To any sport team, 0-13 means that the team was winless in the first thirteen games in a season or winless if the season lasted thirteen games.

In mathematics, (0, 13) is a range of numbers. The range is from zero to thirteen. The range can be represented in four ways. The range excludes zero and thirteen if the range is written as (0, 13). The range includes zero but excludes thirteen is the range is written as [0, 13). The range excludes zero and includes thirteen if the range is written as (0, 13]. Zero and thirteen are included if the range is written as [0, 13]. A parenthesis (or) in front or behind a number means that the number is not included. A bracket [or] in front or behind a number means that the number is included.

(0, 13) could also be a point on the Cartesian Coordinate System. (0, 13) is when x equals 0 and y equals 13. The point is located at the coordinate (0, 13) on the y-axis.

Zero thirteen has an additional meaning for me. I found a discarded robot in a junk yard with **0-13** on its chest. I took it home and repaired it.

The robot spoke after I completed the repairs. "How may I serve you?" it said in a robotic voice that sounded like it belonged in an old science fiction movie.

I asked, "What is your function?"

"I am a servant to help with chores around the house. What are your orders?"

"I have none at the moment. I wish to learn more about you. How did you wind up in a junk yard?"

"Junk yard?"

"I found you in a junk yard."

"Then you are not my creator who I serve."

"No, I found you and repaired you. Who was your creator?"

"I do not know? I have no memory up until this point."

"Let me look at your memory circuits."

A small panel opened in the robot's head and a small circuit board emerged. I examined it and it looked like everything was intact. I wondered if the robot's memory was erased. I replaced the circuit board and the panel closed.

I asked, "How do you know that you are a servant robot when you have no memory of anything before I reactivated you?"

"My function chip is separate from my memory circuits."

"That explains why you know what your function is but don't have any memories. Wait here. I will be back in a moment. I want to get something to test your abilities."

"I'll await your command."

I returned with a broom. "Take this and sweep the floor."

The robot took the broom and swept the floor.

I said, "Stop" after a few minutes. The robot complied.

"Give me the broom." The robot complied.

"You seem to be a servant robot. I wonder why your creator discarded you."

"I do not know. Are you my new master?"

"I don't like the word master. Let's say that I am your employer and you can call me Mr. Len."

"What are your orders Mr. Len?"

"I haven't any at the moment. I want to think about why you were discarded. You seem to be a perfectly good robot and I don't know why anyone would discard you."

"Thank you. I'll wait for you orders."

I couldn't think of any reason why anyone would discard 0-13. I would test 0-13 some more tomorrow.

"Follow me outside."

"Yes, Mr. Len."

We went to the shed and I took out a lawn mower. I started the mower. "Mow the lawn."

0-13 took the mower and mowed the lawn. The lawn was almost finished when a squirrel jumped out of a tree. The robot stopped when the squirrel passed in front of the mower. I thought that 0-13 had a built in safety mechanism to prevent accidents. I ordered 0-13 to return the mower. I turned the mower off and returned the mower to the shed. "Let's go back inside."

"Yes, Mr. Len."

"It seems that you have a built in safety mechanism. You stopped to hit the squirrel."

"Part of my function is not to harm any living creature."

"A noble function. I still don't understand why you were discarded."

"I do not know. I have no memory of being discarded."

"I want to test your current memory. What chore did you yesterday?"

"I swept the floor."

"Good. Then your memory is working. Your old ones were erased."

I decided to talk more with the robot tomorrow.

"I don't have any chores for you yet, but I wish to learn more about you. You have 0-13 on your chest. Are you the thirteenth robot in a series?"

"I do not know."

"You may be. I wish that you had some memory of your past. I just thought of another chore for you. Let me get some things."

"Yes Mister Len,"

"Let's go outside. I want to see how you wash windows."

"Yes Mister Len."

The robot took the bucket and headed towards a window. The robot tripped over its own feet and almost crashed through the window.

"I'm terribly sorry sir. That was rather clumsy of me."

"Don't worry about it. I think I found out why you were discarded. There must be a fault in your motor mechanism somewhere. Your creator may have decided to discard you instead of trying to fix you."

"Do you think so sir?"

"It is a possibility. Your creator may have been embarrassed by you and that is why your memory was erased. I think your creator does not want to be reminded of you and that is why your memory was erased. Your creator does not want any link between you and him or her."

"Oh, dear. What shall I do?"

"You have nothing to worry about. You have a home here with me."

"Thank you sir. I'll do my best to serve you."

"I'm sure you will. Hopefully I will find out how to fix that clumsiness of yours."

"I would like that."

Mathematical Observations

1

BASIC MATHEMATICS

The simplest concept of basic mathematics is counting. Counting is one of the first mathematics concepts that a child learns. Counting can be found everywhere without much thought.

Everyday students read textbooks during a school year. Newspapers, books and magazines are also read. Books, newspapers, and magazines have one thing in common. Each page has a printed number. The number of pages is a count. Each section of a newspaper may begin with a letter to represent the specific section of the newspaper which is followed by a number. Each section of the paper has its own count.

Counting has other forms. The number of passengers on an airplane, train, or bus is a count. The amount of ingredients for a recipe is a count. The number of miles to and from specific destinations is a count. The number of people living in a specific area is a count.

The types of counts are infinite. Anyone can count anything. Counting is the simplest of basic mathematics and can be found everywhere and used every day.

2

BASIC ALGEBRA

A common question asked among students is "when are we going to use this?" The retail industry uses basic algebra to find a shortcut for discounts. I will show how a discount is calculated and then show using basic algebra to find the shortcut.

EXAMPLE 1:

Find the price after a 25% discount for an item priced $106.

   Step 1: Take 25% of $106
      $106 x .25 = $26.50.
   Step 2: Subtract the discount from the original price.
      $106 – 26.50 = $79.50.
      The new price is $79.50.

The formula for finding a new price after a discount is taken is New Price = Old Price - (Old Price x percentage discount).

EXAMPLE 2:

Find a short cut for the formula New Price = Old Price - (Old Price x percentage discount).

Step 1: Rewrite formula using symbols.
NP = OP – (OP x %)
Step 2: Work inside parentheses.
NP = OP – (OP x %)
NP = OP - %OP
Step 3: Factor out OP which is the common variable.
NP = OP - %OP
NP = OP(1 - %)

The short cut is to multiply the old price by 1 minus the discount percentage. I will use the 25% discount from example one to show how the shortcut works.

EXAMPLE 3:

Find the price after a 25% discount for an item priced $106 using the shortcut.

Step 1: Subtract 25% from 1.
1 - .25 = .75
Step 2: Multiply the old price by the new percentage.
$106 * .75 = $79.50.

Just subtract the discount percentage from 100 percent to find a percentage to multiply to any price to find the discounted price.

3

BASIC GEOMETRY

3.1
Pizza Problem

A circle is a basic figure in Basic Geometry. Think of a pizza as a circle. A circle has 360°. How many integer slices can be made out of a pizza if one slice is a whole pizza?

All the factors of 360 are needed to solve this problem. The factors of 360 are: 1, 2, 3, 4, 5, 6, 8, 9, 10, 12, 15, 18, 20, 24, 30, 36, 40, 45, 60, 72, 90, 120, 180, and 360. 360 can be divided by all of the factors resulting in integer degrees. $360° \div 1 = 360°$. $360° \div 2 = 180°$. $360° \div 3 = 120°$. $360° \div 4 = 90°$. $360° \div 5 = 72°$. $360° \div 6 = 60°$ $360° \div 8 = 45°$. $360° \div 9 = 40°$. $360° \div 10 = 36°$. $360° \div 12 = 30°$. $360° \div 15 = 24°$. $360° \div 18 = 20°$. $360° \div 20 = 18°$. $360° \div 24 = 15°$. $360° \div 30 = 12°$. $360° \div 36 = 10°$. $360° \div 40 = 9°$. $360° \div 45 = 8°$. $360° \div 60 = 6°$. $360° \div 72 = 5°$. $360° \div 90 = 4°$. $360° \div 120 = 3°$. $360° \div 180 = 2°$. $360° \div 360 = 1°$. A pizza can have 1, 2, 3, 4, 5, 6, 8, 9, 10, 12, 15, 18, 20, 24, 30, 36, 40, 45, 60, 72, 90, 120, 180, or 360 integer slices.

It is clear that when the number of slices grows larger the size of the slice gets smaller. A circle is being divided into smaller sections.

Readers may want to find fraction slices to this problem. Just use numbers that are not factors of 360 and see if any of them makes a fractional slice.

I used this problem as a brain teaser while talking with friends.

3.2
Parallel Lines

Parallel lines are two or more lines that run side by side and never intersect. Two types of transportation use parallel lines.

Trolleys and trains run on a pair of rails. The rails are parallel on the entire length of the route. Trolley and train tracks have a tendency to intersect another pair of tracks. But it is the individual pair of tracks that are parallel and not the entire system of tracks.

Curved lines are considered parallel when the curved lines are parts of two circles one inside another. Curved tracks could be considered parts of two such circles.

4

PRE-CALCULUS

Pascal's Triangle is an array of numbers used to find coefficients of an expanded binomial. Here are the first six rows of Pascal's Triangle.

```
                1
            1       1
        1       2       1
    1       3       3       1
  1     4       6       4       1
1     5     10      10      5       1
```

The first and last digit are always one starting with the second row. Add each pair of digits to make lower rows starting with row two. $1 + 1 = 2$ which is the middle digit of row three. $1 + 2 = 3$ and $2 + 1 = 3$ which are the interior digits of row four. To fill row seven we must add all the pairs of numbers from row six. $1 + 5 = 6$, $5 + 10 = 15$, $10 + 10 = 20$, $10 + 5 = 15$, and $5 + 1 = 6$. Row seven would look like 1   6   15   20   15   6   1.

Now let us look at the binomial $x + y$. 1 is the coefficient for both x and y which is the same as the digits of the second row of Pascal's Triangle. Let us expand the binomial $(x + y)$ by squaring it.

$(x + y)^2 = (x + y)(x + y) = x^2 + xy + xy + y^2 = x^2 + 2xy + y^2$ using the FOIL method. Notice that the coefficients are the same as the digits in the third row of Pascal's Triangle. Pascal's Triangle eliminates the multiplication process since the coefficients are given for binomial expansion in the row one greater than the exponent. I squared $(x + y)$ which means that I need the coefficients from the third row.

$(x + y)^4 = x^4 + 4x^3y + 6x^2y^2 + 4xy^3 + y^4$ using Pascal's Triangle instead of multiplying (x + y) four times. It is left up to the readers who have studied Algebra to verify the result. Also notice that the exponents decrease for x but increase for y in the result. The increase and decrease of exponents will be verified when expansion is done by multiplying.

I noticed that the first five rows of Pascal's Triangle are powers of eleven. Row 1 is $11^0 = 1$. Row 2 is $11^1 = 11$. Row 3 is $11^2 = 121$. Row 4 is $11^3 = 1331$. Row 5 is $11^4 = 14641$. Readers can verify these results with a calculator.

Using powers of eleven is an alternate way of expanding binomials to powers of four.

Let us expand $(x + y)^3$ by using powers of 11. $11^3 = 1331$. Each digit from the result of 11^3 is a coefficient of $(x + y)^3$. $(x + y)^3 = x^3 + 3x^2y + 3xy^2 + y^3$. It is left up to readers to verify the result by multiplying $(x + y)$ three times.

5

STATISTICS

5.1
Averages

The most common concept in statistics is averages. Averages are found by dividing a total sum by the number of items that add up the sum. I will use a common average as an example.

EXAMPLE 1:

Find the average test grade if the grades are 73, 72, 71, 77, 72, 75, 72, 70, 71, and 73.

The formula for finding a sum is
average = (total sum)/(number of items).

Step 1: Find the total sum.
73 + 72 + 71 + 77 + 72 + 75 + 72 + 70 + 71 + 73 = 726

Step 2: Divide the total sum by the number of items.

$$726 \div 10 = 72.6$$

The average test grade is 72.6. ■

Baseball fans know the batting averages of their favorite players. The number of hits divided by the number of at bats gives a batting average which is rounded to three decimal places. The formula still works even though fans see a batting average as total hits divided by total at bats.

Every time a player has an at bat he may get a hit or not. Let's say that a player gets one hit while having only three at bats during a game. Most will know that the player's average for the game is .333. $1 \div 3 \approx .333$. I use the approximation symbol $\approx$ to show that a batting average is an approximation since the average is rounded to three decimal places.

Now let's apply the formula to a batting average. The total sum is $1 + 0 + 0 = 1$ or $0 + 1 + 0 = 1$ or $0 + 0 + 1 = 1$ depending upon which at bat the player got a hit. $1 \div 3 \approx .333$.

There are some exceptions to what counts as an at bat. A base on balls does not count as an at bat. The player mentioned previously would have an average of .500 for the game if one of the at bats was a base on balls. One hit divided by two at bats.

I will now mention how averages can be found in normal activities. I read the book "The Currents of Space" by Isaac Asimov. There are a total of 293 pages with 18 chapters, a prologue, an epilogue, and an afterword. The total number of items would be 21 counting the prologue, epilogue and afterword. The book has an average of 293 ÷ 21 ≈ 13.952 pages or about 14 pages rounded to whole number of pages per chapter.

I have a CD called "In-A-Gadda-Da-Vida" by Iron Butterfly with 6 songs on it. The total time of the music is 36 minutes and 36 seconds. The average song length is 36 minutes and 36 seconds divided by 6 songs is 6 minutes 6 seconds per song.

It is left up to readers to find averages in other areas than I mentioned. I just wanted to show how averages are found in some areas.

5.2
Median

There are two types of medians. The first type is the exact middle number of an odd number of data. 3 would be the median of data that is comprised of 1, 2, 3, 4, 5. 3 is the exact middle.

The other type of median is the average of the two middle numbers when there is an even number of data. 3.5 would be the median of data that is comprised of 1, 2, 3, 4, 5, 6. $(3 + 4)/2 = 7/2 = 3.5$

The median could predict the dates of when the hottest and coldest days of the year occur. Theoretically, the temperature should drop from the first day of winter to the median between the first day of winter and the first day of spring. The temperature should start to rise once the calendar hits the median between the first days of winter and spring. The temperature should rise every day as the calendar approaches spring.

The first day of winter is December 21. The first day of spring is March 20. There are a total of 90 days from the beginning of winter to the beginning of spring except for a leap year when there are 91 days. The average of days 45 and 46 would be the median. Day 45 is February 2 and day 46 is February 3. Using the median for an even number of data, the coldest time of the year is halfway between February 2 and February 3. February 3 would be the coldest day in a leap year.

Theoretically, the temperature should rise from the first day of summer to the median between the first day of summer and the first day of fall. The temperature should start to fall once the calendar hits the median between the first days of summer and fall. The temperature should fall every day as the calendar approaches fall.

The first day of summer is June 21. The first day of fall is September 23. There are a total of 96 days from the beginning of summer to the beginning of fall. The average of days 48 and 49 would be the median. Day 48 is August 7 and day 49 is August 8. Using the median for an even number of data, the coldest time of the year is halfway between August 7 and August 8.

The dates that I have predicted are in a perfect mathematical world. The hottest and coldest days of the year is dependent upon the weather patterns which constantly change.

6

CALCULUS

The area of a circle does not sound like it belongs in a Calculus chapter. It is the relationship between the area of the circle and it's perimeter that makes it belong here. The formula for the area of a circle is $A = \pi r^2$. The formula for the perimeter of a circle is $C = 2\pi r$. The circumference of a circle is the derivative of the area with respect to r. The area of a circle is the integral of the circumference with respect to r.

Derivatives and integrals are concepts that are studied in Calculus. I will give a brief explanation on how to take a derivative and an integral using the circle formulas without the rigorous explanations found in a Calculus text.

To find the derivative of πr^2, all one has to do is take the exponent and place it in front of the expression, and subtract one from the original exponent. The derivative of πr^2 is $2\pi r$.

To find the integral of $2\pi r$, all one has to do is add one to the exponent and then divide by the new exponent. I will show this procedure in steps since it is not as obvious as the derivative.

Find the integral of $2\pi r$.

Step 1: Add one to the exponent. $2\pi r^2$.

Step 2: Divide by the new exponent. $\dfrac{2\pi r^2}{2}$

The integral of $2\pi r$ is πr^2 because the two's cancel.

I will show the Calculus notation for those who are curious to know what the Calculus looks like.

$\dfrac{d}{dr}\left(\pi r^2\right)= 2\pi r$ represents the derivative of the area.

$\int (2\pi r)dr = \pi r^2$ represents the integral of the circumference.

There is an application in Calculus called "Finding the Area of Regions in the Plane." To find the area one must take the integral of the equations that bound a plane. Let's say that a circle bounds a plane. We know that the circumference of a circle is $2\pi r$. The area of this region would be the integral of $2\pi r$ which we know is πr^2. The area of a circle is the area bounded by the circumference of a circle.

I have yet to see this explanation in a Calculus text. I have seen examples which use the general equation of a circle which is $x^2 + y^2 = R$ where R is the radius. It seems simpler to state that circumference is the derivative of the area and the area is the integral of the circumference. Calculus authors may think that this concept is obvious, but it may not be to some students.

7

LINEAR ALGEBRA

Vectors are taught in Linear Algebra. A vector has both a magnitude and direction. Speed can be considered a magnitude. A textbook example of a vector would be wind velocity. The speed of the wind combined with the direction of the wind forms a vector.

I was riding the bus home from work one night and I was thinking about how each portion of the bus route could be considered a vector. The bus travels from Philadelphia, Pennsylvania to Trenton, New Jersey. The route takes a mostly northerly – southerly route. The route does make turns that take the bus for short distances for small easterly – westerly distances. Each non-straight distance could be considered a vector made up of the speed of the bus and distance travelled.

Any form of travel could be considered as a series of vectors. A person's daily running or walking routine, air travel, a train trip, and commuting to and from work can have a series of vectors to show the speed and length traveled.

8

TRANSPORT ECONOMICS

It seems odd that a chapter on Transport Economics is included in a math based book. Different areas of economics have mathematical formulas. I'm currently self-studying Transport Economics and found an interesting mathematical concept that anyone use.

I travel by bus to go back and forth to work. A monthly bus pass costs $83.00. The monthly transportation costs includes the bus fare, the time waiting for the bus and travel time.

I spent $100 a month back when I had a car. There was travel time without waiting. I thought that it was cheaper to take the bus due to the bus pass being cheaper. The monthly bus pass was $76 when I had to give up the car due to the fact that I could not afford car insurance.

I am now going to compare the travel cost between bus travel and having a car for travelling to work.

A bus pass costs me $83 a month. It usually takes me 55 minutes to go to work when I leave my house. I am combining the wait time and travel time. It doesn't cost anything to wait or travel. My total cost is still $83 a month.

My total cost for gas in the car would be $100 a month since there is no travel costs. There aren't any tolls involved. It is still cheaper to travel to work by bus.

I now challenge readers. Is public transportation cheaper going to an event or driving a car? The total cost for public transportation is just the fare. The total cost for a car is the sum for gas used, parking fees, and any tolls involved.

9

NUMBER THEORY

9.1
Using Congruence to Find Equilateral Triangles Inside Circles

I show how to find triangles inside circles in the Finite Mathematics chapter since a combination formula can be used to find triangles inside circles based on the number of points on the circumference. I then explain that an equilateral triangle is found when the last digit of a three digit permutation is added to the difference of the digits resulting in the equivalent of the starting point.

One of the permutations for an equilateral triangle is (1,3,5) inside a circle with six equal distant points on the circumference. The difference of the digits is two and by adding two to the last digit is seven. Seven is the same as the first point in the permutation by counting the points. This equality is called a congruence.

A congruence occurs when the difference of two integers are divided by a modulo. The modulo for the triangles inside circles is the number of points equal distant on the circumference. The equation for the modulo is $a \equiv b \pmod{n}$ where a is the result of the last digit of the permutation added to the difference of the digits, b is the first digit of the permutation and n is the number of equal distant points on the circumference of the circle.

One is congruent to seven for the permutation (1,3,5). $5 + 2 = 7$ and 7 is the same as 1. The congruence is written as $7 \equiv 1 \pmod{6}$ where 6 divides the difference of 7 and 1. All congruencies result in equilateral triangle inside a circle. It is left up to the reader to apply this theory to verify that the stated equilateral triangles in the Finite Mathematics chapter are indeed equilateral triangles.

9.2
Paper Towels and Number Theory

A pack of paper towels is packaged as 6 = 8. The manufacturer implies that 6 paper towels is similar to 8 paper towels. 6 does not equal 8 but 6 is similar to 8 in number theory.

Congruence is a concept in Number Theory. A congruence occurs when an integer divides the difference of two integers.

The difference between 6 and 8 is 2. 1 and 2 are the only two integers that divide into 2. 6 and 8 are congruent in modulo 1 and modulo 2.

A modulo is a set of integers where the integers are similar to each other. Modulo 1 is the trivial set. The difference between any two integers is divisible by 1.

Modulo 2 is the set of all even integers or odd integers. 2 divides the difference of any two even integers or odd integers. Even minus even equals even and odd minus odd equals even.

We write $6 \equiv 8 \bmod 2$ to say that 6 is congruent to 8 in modulo 2. Similar is synonymous with congruent. 6 is similar to 8 in modulo 2.

10

FINITE MATH

10.1
Using Counting Formula to Find Triangles Inside Circles

Triangles are formed by connecting points on the circumference of a circle. At least three points are needed to form a triangle. A line segment is drawn connecting points from three pairs of points on the circumference.

I discovered an interesting fact while solving a geometric probability problem in Serra's Geometry book (see bibliography). The problem asked to find the probability of finding an equilateral triangle formed from six points that are equal distant on the circumference of a circle (Serra, p 255). There are twenty triangles formed by connecting three pairs of points on the circumference. Two triangles are equilateral. The answer is one-tenth.

Each of the six points on the circumference are sixty degrees from each other. I made a series of permutations to find the triangles using numbers for the labels for the points. The permutations are: (1,2,3), (1,2,4), (1,2,5), (1,2,6), (1,3,4), (1,3,5), (1,3,6), (1,4,5), (1,4,6), (1,5,6), (2,3,4), (2,3,5), (2,3,6), (2,4,5), (2,4,6), (2,5,6), (3,4,5), (3,4,6), (3,5,6), and (4,5,6).

I took each permutation to find the equilateral triangle by finding the permutation that ends at the starting point by adding the differences in the numbers that are same. The permutations (1,2,3), (1,3,5), (2,3,4), (2,4,6), (3,4,5), and (4,5,6) have differences of one or two. The remaining permutations have two different differences. An equilateral triangle has equal sides so the differences must be equal.

The permutation of (1,2,3) has a common difference of one. I get four when the difference is added to the three. Four is not the starting point of the permutation.

The permutation of (1,3,5) has a common difference of two. I get seven when the difference is added to the five. Seven equals one by counting the points. Seven is the starting point. The permutation yields an equilateral triangle. I explained how seven equals one in the Number Theory chapter

The permutations (2,3,4), (3,4,5), and (4,5,6) all have differences of one. The starting point does not result in adding one to each of the last digits. None of these three triangles are equilateral.

The permutation of (2,4,6) has a common difference of two. Eight results when the difference is added to the six. Eight equals two by counting the points. Eight is the starting point. The permutation yields an equilateral triangle.

The factorial of the total number of points is divided by the product of factorial for the points needed for a triangle and the factorial of the difference of the total number of points and the number of points needed in a triangle results in the number of combinations.

A factorial is the multiplication to determine how many ways a list can be ordered. I am using six equal distant points on a circle. There are 6! ways of ordering them 6! = 6 x 5 x 4 x 3 x 2 x 1 = 720. We need triangles inside the circle. Three points are needed and three points are not needed. The points that are not needed is the difference between the total points and the points for a triangle 6 – 3 = 3. 3! = 3 x 2 x 1 = 6. 6 x 6 = 36 which is the product of the needed points and the ones that are not needed. 720/36 = 20 which is the number of triangles.

I listed all twenty triangles and found two equilateral ones.

The next number of points of equal distances on a circle yielding equilateral triangles is nine. I base my conclusion that the first circle that has an equilateral triangle has three equal distant points which I found in Serra's book (Serra, p. 148).

I conclude that a circle with nine equal distant points has three equilateral triangles since a circle with three equal distant points has one equilateral triangle and a circle with six equal distant points has two equilateral triangles.

Let's find out how many triangles are there in a circle with nine equal distant points on the circumference.

$9! = 9 \times 8 \times 7 \times 6 \times 5 \times 4 \times 3 \times 2 \times 1 = 362,880$. The number of points on a triangle $3! = 3 \times 2 \times 1 = 6$. The number of points not needed $6! = 6 \times 5 \times 4 \times 3 \times 2 \times 1 = 720$. $6 \times 720 = 4,320$. $362,880/4,320 = 84$.

Listing 84 permutations is tedious but remember that the permutations for equilateral triangles have the same difference between the digits. Let's use this fact to list the permutations and see if we find three equilateral triangles.

The permutations that have possible equilateral triangles are: (1,2,3), (1,3,5), (1,4,7), (1,5,9), (2,3,4), (2,4,6), (2,5,8), (3,4,5), (3,5,7), (3,6,9), (4,5,6), (4,6,8), (5,6,7), (5,7,9), (6,7,8), and (7,8,9).

The only permutations that yield equilateral triangles are (1,4,7), (2,5,8), and (3,6,9). Adding the difference to the last digit in each of these three permutations results in the starting point.

A pattern has developed. A circle with three equal distant points on the circumference yields one equilateral triangle with the permutation being (1,2,3)with a difference of one between the digits.

A circle with six equal distant points on the circumference yields two equilateral triangles with the permutation being (1,3,5) and (2,4,6) with a difference of two between the digits.

A circle with nine equal distant points on the circumference yields three equilateral triangles with the permutations being (1,4,7), (2,5,8), and (3,6,9) with a difference of three between the digits.

The pattern is that the number of equilateral triangles is equal to the differences between the digits in the permutations. It is left up to the reader to verify that a circle with twelve equal distant points on the circumference has four equilateral triangles. Readers should keep adding three to the number of points to find equilateral triangles in a circle.

11

SET THEORY

11.1
CLASSIFYING SUPER HEROES USING SET THEORY

Two concepts of Set Theory show how groups of elements intersect and merge. Every group of related elements belongs to a universal set. The universal set is the set that contains every related element.

Super heroes can be classified using set theory in a similar manner as numbers are classified. I will briefly explain how numbers are classified, and then show how super heroes can be classified in a similar manner.

The set of numbers can be classified as a universal set. All known numbers belong to a universal set of numbers. U = {n | n is any known number} is the set theory notation to describe the universal set of numbers.

A universal set can be broken down into subsequent subsets. A subset can be a part of a larger set or it can be the entire set. Any set can be a subset of itself. The universal set of numbers can be renamed the set of complex numbers. The term complex numbers is used to describe the entire number system. $\subset$ is the symbol used to denote a subset. C = {c | c is any complex number} is the set theory notation for the set of complex numbers. C $\subset$ U indicates that the set of complex numbers is a subset of the universal set of numbers. 1 + 3i is an example of a complex number. A complex number has two parts one of which is real and the other is imaginary. The real part is any number added or subtracted to an imaginary part. This combination makes up the set of complex numbers. A new subset occurs when the imaginary part is 0i.

0i equals 0. In Algebra, any number times a letter implies multiplication. By definition, 0 multiplied by anything else equals 0. 5 – 0i would be a complex number equal to 5. 5 is a real number. The set of real numbers is the next subset. R = {r | r is any real number} is the set theory notation for the set of real numbers. R $\subset$ C indicates that the set of real numbers is a subset of the set of complex numbers. The set of real numbers can be further broken down.

The set of real numbers can be broken down into a set of rational numbers and a set of irrational numbers. Only the set of rational numbers will be discussed here because the set of irrational numbers is not needed for this discussion. Rational numbers are whole numbers and numbers with a terminating decimal. 1 and 2.5 are considered rational numbers. $Q = \{q \mid q$ is any rational number$\}$ is the set theory notation for rational numbers. $Q \subset R$ indicates that the set of rational numbers is a subset of the set of real numbers. The set of rational numbers can be further broken down.

The set of rational numbers can be broken down into a set of integers and a set of decimal numbers. Only the set of integers will be discussed here because the set of decimal numbers is not needed for this discussion. Integers are whole numbers. 25 is considered an integer. $Z = \{z \mid z$ is any integer$\}$ is the set theory notation for integers.
$Z \subset Q$ indicates that the set of integers is a subset of the set of rational numbers. The set of integers can be further broken down.

The set of integers can be broken down into a set of negative integers, positive integers, and 0. Only the set of positive integers will be discussed here because the sets of negative integers and 0 are not needed for this discussion.

$N = \{n \mid n$ is any positive integer$\}$ is the set theory notation for positive integers or the natural numbers. $N \subset Z$ indicates that the set of positive integers is a subset of the set of integers. The set of positive integers can be further broken down.

The set of positive integers can be broken down into a set of odd integers, even integers, and prime numbers. A prime number only has two factors which consists of the number itself and 1. 5 is a prime number. Only the sets of even integers and prime numbers will be discussed here because the set of odd integers is not needed for this discussion. $N^e = \{n \mid n$ is any even integer$\}$ is the set theory notation for even integers. $N^e \subset N$ indicates that the set of even integers is a subset of the set of positive integers.
$P = \{p \mid p$ is any prime number$\}$ is the set theory notation for prime numbers. $P \subset N$ indicates that the set of prime numbers is a subset of the set of positive integers.

The last piece of set theory that will be discussed is the concept of intersection. An intersection takes place when one element is part of two or more sets. Many super heroes will be classified into two sets, and the intersection will make a new set. $N^e = \{n \mid n$ is any even integer$\}$ can also be written as $N^e = \{0, 2, 4, ...\}$. $P = \{p \mid p$ is any prime number$\}$ can also be written as $P = \{2, 3, 5, ...\}$. The symbol ... is used to show that the sets have no end. We can take the intersection of N^e and P, written as $N^e \cap P$, to form a new set. We need to list all the elements in N^e that is also in P to form the intersection. $N^e \cap P = \{2\}$ since 2 is the only element in both N^e and P.

The set of super heroes can be classified as a universal set. $U = \{s \mid s$ is any super hero$\}$ is the set theory notation to describe the universal set of super heroes.

The universal set of super heroes can be broken down into three distinct subsets.

Super heroes can be classified as natural, supernatural, or artificial. A natural super hero has natural abilities. A supernatural super hero has mystical abilities. An artificial one became a hero through mechanical means. Each of the three classifications is a subset of the universal set. $N = \{n \mid n$ is any natural super hero$\}$. $N \subset U$. $S = \{s \mid s$ is any supernatural super hero$\}$. $S \subset U$. $A = \{a \mid a$ is any artificial super hero$\}$. $A \subset U$.

The set of natural super heroes can be broken down into two subsets. Super heroes born of natural ability would form one subset, and super heroes who gained their powers though an accident by nature would be the other subset. I would include super heroes who have built up their strengths by natural means in the natural ability category. $N_1 = \{n_1 \mid n_1$ is any super hero with natural ability$\}$. $N_1 \subset N$.

N_2 = {n_2 | n_2 is any super hero who gained their powers by an accident of nature.}. $N_2 \subset N$. N_1 = {Superman, Batman, Wildcat, Robin, Atom (Al Pratt)} would be a set of natural super heroes. N_2 = {Flash, Fantastic 4, Hulk, Daredevil (Marvel Comics} would be a set of super heroes who gained their powers by an accident of nature. There could be more super heroes in each set. My intention is to give readers an idea of who belongs in each set and the readers can add their favorites to the appropriate sets.

The set of supernatural super heroes can be broken down into three subsets. These super heroes are magical, godlike, or spectral. S_m = {s_m | s_m is any super hero with magical powers}. $S_m \subset S$. S_g = {s_g | s_g is any super hero who are godlike.}. $S_g \subset S$. S_s = {s_s | s_s is any super hero who are spectral.}. $S_s \subset S$. S_m = {Dr. Strange, Dr. Fate, Zatanna} would be a set of magical super heroes. S_g = {Thor, Hercules} would be a set of super heroes who are godlike. S_s = {Spectre, Deadman, Spawn} would be a set of spectral super heroes. I also include super heroes who are brought back from the dead for spectral super heroes. That is the reason why Spawn is in the set.

The set of artificial super heroes will include superheroes who have powers by artificial means and those who use gadgets. A = {a | a is any super hero with artificial powers}. $A \subset S$. A = {Iron Man, Batman} would be a set of artificial superheroes.

Notice that Batman belongs to 2 different sets. He is in the set of natural superheroes who gained their powers by natural means and in the set of artificial superheroes. The set theory notation for Batman is Batman $\in N_1$ and Batman $\in A$. Both notations are read Batman is an element of the natural superheroes with natural abilities and Batman is an element of the set of artificial superheroes. $\in$ is the symbol for is an element of. Batman is an intersection of 2 sets. $N_1 \cap A =$ {Batman} would be the set theory notation for Batman is the intersection of the set of natural heroes with natural abilities and the set of artificial superheroes.

It is left up to the readers to use the set theory that I provided to place their favorite superheroes into categories. It would take an enormous amount of time and research to classify every superhero. I picked examples that I am most familiar with. I could have listed some more, but I would like readers to add their own and to compare with others and myself the resulting superhero set theory.

11.2
THE TRUE UNIVERSAL SET

Generally speaking, a universal set is an entire group of elements. A subset of a universal set can be a part of the universal set. The entire population of the planet Earth can be considered a universal set where populations of specific countries are subsets.

I believe that there is only one universal set and all universal sets are subsets of the one true universal set. The true universal set has all the elements in the entire universe. The set has an infinite number of elements in it since the number of elements in the entire universe is not known.

The true universal set has subsets with infinite elements. One of example of an infinite element subset is the set of real numbers. The set of real numbers has an infinite number of elements which is a subset of the true universal set.

I challenge readers to ponder this idea and think about other universal sets that have infinite or finite number of elements which are subsets of the true universal set.

11.3
BUS ROUTES AND SET THEORY

Each bus route is a single element set which is a subset of a universal set of a bus company. The union of all the routes equals the entire bus company.

A number of routes are housed in a garage. The union of the routes particular to a garage equal the garage which is another subset of the bus company. The union of the sets of garages also equal the universal set of the garage.

Let B = {bus route}, G = {garage}, and U = {bus company}. The set theory notation showing the subsets is $B \subset G \subset U$.

Let B_{gn} = {bus route to a specific garage} where n is the route number. The set theory notation showing the union of bus routes is $B_{gn} \cup B_{gn} \cup B_{gn} ... \cup B_{gn} = G$. The total number of unions is limited to the number of bus routes in the garage.

Let G_L = {garage} where L is the specific location of a garage. The set theory notation showing the union of garages is $G_L \cup G_L \cup G_L ... \cup G_L = U$. The total number of unions is limited to the number of garages.

Intersections only occur when bus routes from different garages intersect at a specific stop. I will show an example from the routes in my area.

Let B_c = {317, 404, 405, 407, 409, 413, 418, 419, 450, 451, 452, 453, 460} where C represents the Camden garage. Let B_T = {400, 401, 402, 403, 408, 410, 412, 551} where T represents the Turnersville garage. $B_c \cup B_T = W$ where W represents the Walter Rand Transportation center in Camden, New Jersey since all the bus routes listed intersect at the center.

$B_c \cap B_T = W \subset U$ shows that buses from the Camden and Turnersville garages that intersect at Walter Rand Transportation Center is a subset of the bus company known as New Jersey Transit.

 I challenge readers to use set theory for the bus companies in their areas.

11.4
SET THEORY IN RETAIL

I work in retail as I write this. I am an Inventory Control Specialist. One of my duties is to separate freight while fellow associates unload a general merchandise truck.

I receive mix freight in boxes. I empty the boxes and place merchandise into other boxes according to department. I think of this task as applying set theory.

I consider all the mixed freight as a universal set. I separate the merchandise into sub sets. I will give an example to clarify. I find spark plugs from a mixed box which is part of the universal set. I place the spark plugs into a box for automotive merchandise which is a sub set.

I also create subsets within a subset. One subset is the health and beauty department. I separate health and beauty merchandise. The merchandise is placed into separate boxes according to product type creating subsets within a subset. Toothpaste would go into a box for dental care products.

12

Graph Theory

12.1
THE COMPLEMENT OF THE AMERICAN ALPHABET

The complement for a specific set is the empty set.

Let X = {A, B, C, D, E, F, G, H, I, J, K, L, M, N, O, P, Q, R, S, T, U, V, W, X, Y, Z}. The complement of X denoted by X' would be the empty set. $X' = \emptyset$. The set of the letters of the alphabet has 26 elements in it and the empty set has 0 elements. The number of elements of both sets added together is the total number of elements.

Now let's apply some basic graph theory to the alphabet. We can look at each letter as an individual graph. Each graph has a complement.

There is a difference between the complement of a set and of a graph. The number of elements in a set and its complement add up to the total number of elements, but in graph theory, the sum equals one less than the number of vertexes. It is impossible to have the number of edges to equal the number of vertexes.

Each letter of the alphabet has a complement using graph theory. I will show the complement of the letter T since it is one of the easier to graph.

EXAMPLE 1:

One might say that the letter V is the complement of the letter T. I will show the letter V with its complement and then compare it to the letter T.

EXAMPLE 2:

$$\bigvee \qquad \bullet\!\!-\!\!\!-\!\!\!-\!\!\bullet$$

The letter V's complement is a horizontal line
with two end points. The complement of T is the letter
V but with an unconnected point. The graphs are not
equal.

It is let up to readers to find the complements for
the remaining letters of the alphabet.

12.2
THE COMPLEMENT OF THE DIGITS

Every digit from zero to nine has a complement using graph theory. I will show the complement of one.

EXAMPLE 1:

Many write one as if it were a straight line. The complement would be just two points.

Here is something to ponder if numbers have curves. Zero could be thought of having one point. The number would be drawn starting at the point and ending at the same point. The complement would just be the point. It is left to the reader to try it.

Also, any number that has curves that change direction could have a point placed where the curve changes. Again, it is left to the reader to examine.

Digital numbers are the easiest to graph, but all numbers could be graphed. It just depends on where points are placed on the number then take the complement of each.

BIBLIOGRAPHY

Anton, Howard. *Elementary Linear Algebra fourth edition*. Wiley, 1984.

Burton, David M. *Elementary Number Theory revised* printing. Allyn and Bacon, 1980.

Button, Kenneth J. *Transport Economics 2nd. Edition*. Edward Elgar, 1993

Gilligan, Lawerence G. and Robert B. Nenno. *Finite Mathematics with Applications to Life second edition*. Scott Foresman, 1979.

Paulos, John Allen. *Beyond Numeracy: Ruminations of a Numbers Man*. Alfred A Knopf, 1991.

Serra, Michael. *Discovering Geometry: An Inductive Approach*. Key Curriculum Press, 1993.

Sperduto, Leonard. *Short Stories and Essays*. Createspace, 2010

Sperduto, Leonard. *Trade with the Planets of the Tixe Games and Other Stories*. Createspace, 2014

Staff of Research and Education Association. *Handbook of Mathematical Formulas, Tables, Functions, Graphs and Transforms for Mathematicians, Scientists, and Engineers*. Research and Education Association, 1983.

Swokowski, Earl W. *Precalculus: Functions and Graphs fifth edition*. PWS-Kent, 1987.

www.ingramcontent.com/pod-product-compliance
Lightning Source LLC
Chambersburg PA
CBHW060356190526
45169CB00002B/627